MathStart™

SUBTRACTING

Elevator Magic

by Stuart J. Murphy ★ illustrated by G. Brian Karas

HarperCollins Publishers

LEVEL 2

To C.L.T. —
who helped me to develop a magical career.
— S.J.M.

For Ben—
who sees things in a different way.
— G.B.K.

HarperCollins®, 🏭®, and MathStart™ are trademarks of HarperCollins Publishers Inc.

For more information about the MathStart Series, please write to
HarperCollins Children's Books, 10 East 53rd Street, New York, NY 10022,
or visit our web site at http://www.harperchildrens.com.

Bugs incorporated in the MathStart series design were painted by Jon Buller.

Elevator Magic
Text copyright © 1997 by Stuart J. Murphy
Illustrations copyright © 1997 by G. Brian Karas
Printed in the U.S.A. All rights reserved.

Library of Congress Cataloging-in-Publication Data
Murphy, Stuart J., date
    Elevator magic / by Stuart J. Murphy ; illustrated by G. Brian Karas.
        p.   cm. — (MathStart)
    "Level 2, subtracting."
    Summary: Explains the concept of subtraction through a rhyming text
about a descending elevator.
    ISBN 0-06-026774-7. — ISBN 0-06-446709-0 (pbk.)
    ISBN 0-06-026775-5 (lib. bdg.)
    1. Subtraction—Juvenile literature.   [1. Subtraction.]   I. Karas, G. Brian,
ill.   II. Title.   III. Series.
QA115.M868   1997   96-5672
513.2'12—dc20   CIP
                                                                              AC

Typography by Elynn Cohen
1   2   3   4   5   6   7   8   9   10
❖
First Edition

Hi, Ben. I'm glad you came up to meet me. I'm all set to go, but we have to make a few stops on the way down.

Is it okay if I push the buttons?

Sure! First we have to cash a check at Farm Bank and Trust. It's 2 floors down from here.

6

Mom said 2 floors down. Which one do I press?

We're on 10.

2 floors down

10-2=8

Farm Bank and Trust

I think that floor 8 would be the best guess.

We're almost there.
We'll be getting off now.

I think I hear oinking and mooing and . . .

In this bank there's a horse,
and chickens, of course,
a donkey, a pig, and even a cow.

Next we have to drop off this package at *Speedway Delivery*. It's 3 floors down from here.

What floor should I push to go down just 3?

Now we're on 8.
3 floors down
8−3=5

Speedway Delivery

Floor 5 is where Speedway Delivery should be.

We're finally here. That seemed kind of slow.

Now I hear rumbling and screeching and . . .

There will soon be a race
in this big, noisy place.
The cars and the trucks are all ready to go.

17

Let's make a special stop at the Hard Rock Candy Store. It's 1 floor down from here.

I'll press the next button to go down 1 floor.

Now we're on 5.

1 floor down

5−1=4

Hard Rock Candy Store

There'll be lots of treats when we get to floor 4.

The door's about to open—I can't wait to see.

The store's filled with sound.
Bright lights spin around.
A rock band is starting to play just for me.

Here's your treat. Now we're all set to go.

We're going to meet Dad, who's on the first floor.

We're on 4.

Dad's on 1.

4−1=3 floors down

Lobby

From floor 4 to floor 1 is 3 floors more.

We're almost there now. Then we'll be on our way.

Now I hear honking and beeping and . . .

HEY!

What a great time I've had.
I can't wait to tell Dad.

My elevator ride was like magic today.

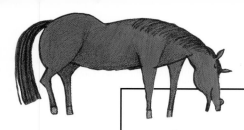

I f you would like to have more fun with the math concepts presented in *Elevator Magic*, here are a few suggestions:

- Read the story together and ask the child to describe what is going on in each picture. Ask: "Which floor would you like to visit?" "Why?"

- Ask questions throughout the story, such as "Which floor is 2 floors down from the 10th floor?" and "If you go 3 floors down from the 8th floor, where will you be?"

- Look at the pictures of the boy pushing the elevator buttons. On each page pick a different number of floors to go down. Then do the subtraction.

- Make your own set of elevator buttons on a piece of paper or cardboard. Start at the top. If you want to go down 2 floors, which button should you push? What if you want to go down 5 floors?

- Next time you are on an elevator, practice subtraction during the ride.

- Look at things in the real world and solve subtraction problems when they occur. For example: If you buy 6 apples and eat 3, how many apples are left? If you have a book of 10 stickers, and give 2 to a friend, how many stickers will you still have?

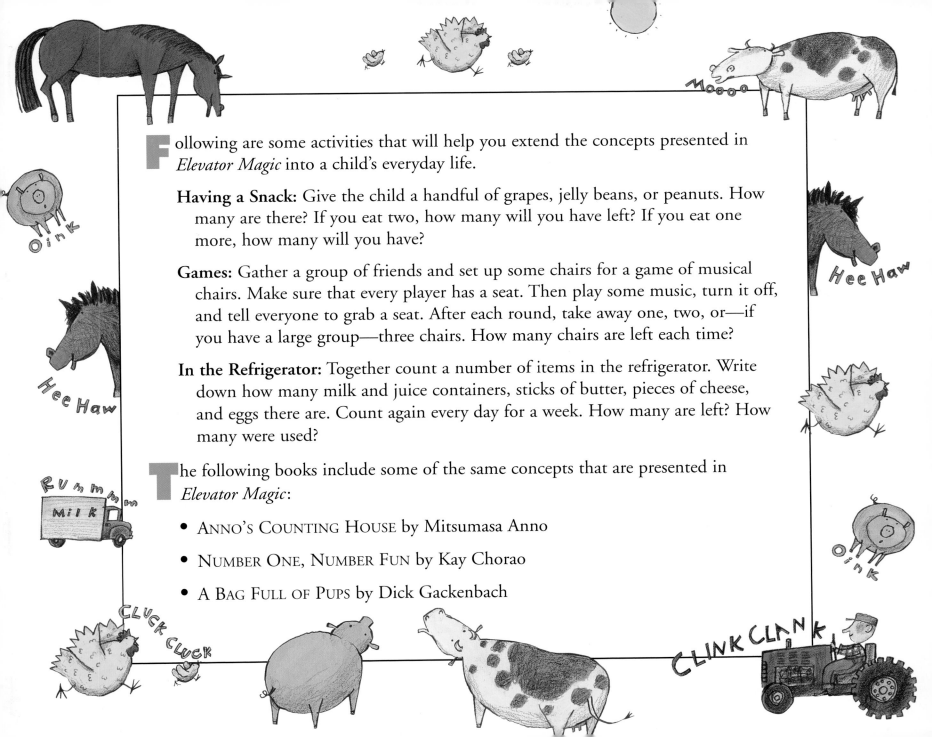

Following are some activities that will help you extend the concepts presented in *Elevator Magic* into a child's everyday life.

Having a Snack: Give the child a handful of grapes, jelly beans, or peanuts. How many are there? If you eat two, how many will you have left? If you eat one more, how many will you have?

Games: Gather a group of friends and set up some chairs for a game of musical chairs. Make sure that every player has a seat. Then play some music, turn it off, and tell everyone to grab a seat. After each round, take away one, two, or—if you have a large group—three chairs. How many chairs are left each time?

In the Refrigerator: Together count a number of items in the refrigerator. Write down how many milk and juice containers, sticks of butter, pieces of cheese, and eggs there are. Count again every day for a week. How many are left? How many were used?

The following books include some of the same concepts that are presented in *Elevator Magic*:

- ANNO'S COUNTING HOUSE by Mitsumasa Anno

- NUMBER ONE, NUMBER FUN by Kay Chorao

- A BAG FULL OF PUPS by Dick Gackenbach